Food 224

我做的是好事!

I Do Good!

Gunter Pauli

[比]冈特·鲍利 著

[哥伦]凯瑟琳娜·巴赫 绘

颜莹莹 译

特别感谢以下热心人士对童书工作的支持：

匡志强　方　芳　宋小华　解　东　厉　云　李　婧
刘　丹　熊彩虹　罗淑怡　旷　婉　杨　荣　刘学振
何圣霖　王必斗　潘林平　熊志强　廖清州　谭燕宁
王　征　白　纯　张林霞　寿颖慧　罗　佳　傅　俊
胡海朋　白永喆　韦小宏　李　杰　欧　亮

目录

Contents

一只猕猴独自坐在一棵棕榈树顶端，双手抱头。一头侏儒象注意到它脸上的表情非常悲伤，问道：

“你怎么了？是你的父母去世了吗？”

“没有那么糟，不过差不多……”

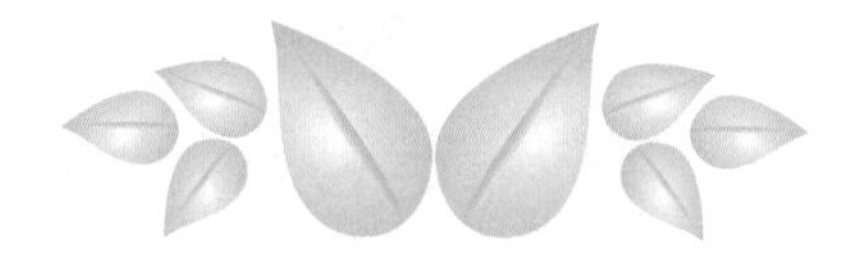

A macaque is sitting alone in the top of a palm tree with his head in his hands. A pygmy elephant notices the very sad expression on his face and asks:

“What happened to you? Did your mom or dad pass away?”

“Nothing quite that bad, but almost ...”

一只猕猴独自坐着……

A macaque is sitting alone ...

我们的大片雨林在20年前就被夷为平地了……

Our large rainforest was destroyed twenty years ago …

“哦，不！不要告诉我他们又被诱捕并抓走了，不要告诉我你们又失去了家园。”

“你知道的，为了给棕榈种植园腾出地方，我们的大片雨林在20年前就被夷为平地了。我们必须学会在剩下的小片森林中生存。”

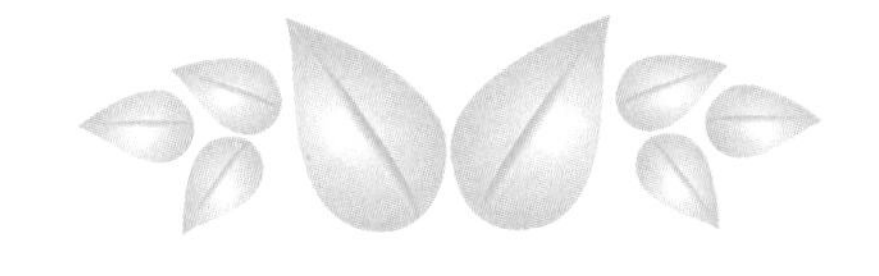

“Oh, no! Don’t tell me they have been trapped and taken away again, or that you are once again losing your home.”

“As you know, our large rainforest was destroyed twenty years ago, burnt to the ground to make room for palm oil plantations. We had to learn to survive in the small pockets of forest that were left.”

“那么，除了孤独，你还感到饥饿吗？”

“一点也不。环顾四周，你会发现有充足的食物。我的家庭也很兴旺。”

“那么请告诉我，如果没有人去世，你的家还在，而且你也不饿，那你为什么愁容满面呢？”

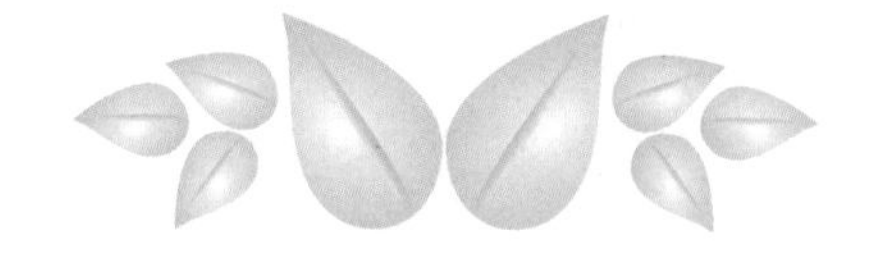

“So, apart from feeling lonely, you are also hungry?”

“Not at all. Look around, and you will see there’s plenty of food. My family is thriving.”

“So tell me then, if no one has passed away, your home is still standing, and you are not hungry, what justifies this sad face?”

……有充足的食物。

... there's plenty of food.

刚才有人说我是害兽。

just been told I am a pest.

“唉，刚才有人说我是害兽，现在他们把我当害兽看待！”猕猴说。

“害兽？你们毁坏庄稼，抢夺钱财和食物吗？”

“几乎没有。首先，是他们夺走我们的土地，我们因而失去了家园和食物。然后他们种植油棕榈，但是我们可以吃一些果实吗？不行！”

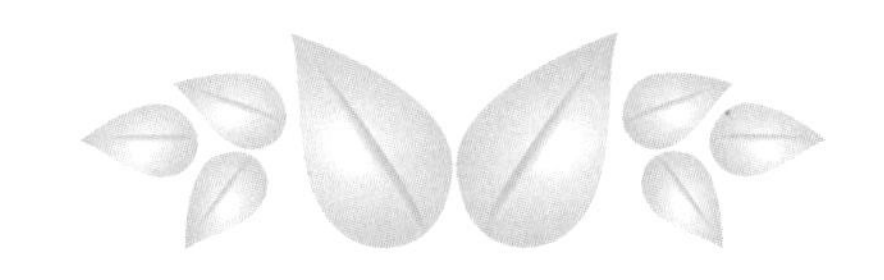

“Well, I have just been told I am a pest, and am now treated as one!” Macaque says.

“A pest? Do you destroy crops and deprive people of money, or food?”

“Hardly. Look, first they took our land, so we lost our homes and food. Then they planted oil palms, but are we allowed to eat some of the fruit? No!”

“我理解你的挫折感。但是请告诉我，你们这些猕猴不是会偷袭油棕榈抢果子吃吗？”

“不！我们有足够的水果、浆果和树叶充饥。我们还会在里面加点肉。”

“那么你就不是我这样的食草动物了？”大象问道。

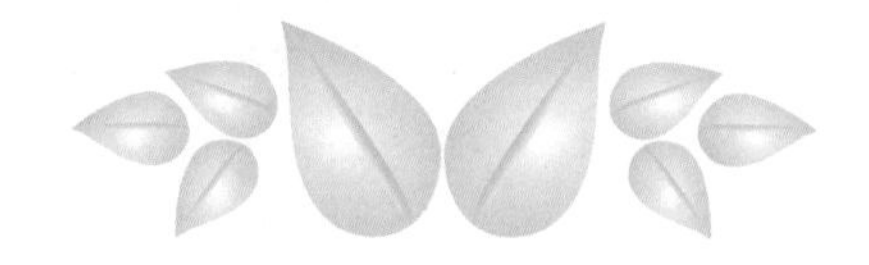

“I do understand your frustration. But tell me, have you macaques not been raiding the oil palms for the dates?”

“No! We have enough fruit, berries, and leaves to eat, to which we add a little meat.”

“You’re not a herbivore then, like me?” Elephant asks.

你就不是我这样的食草动物了？

You're not a herbivore like me?

老鼠？你吃老鼠？

Rats? You eat rats?

“不，我们更像是杂食动物。我确实喜欢吃点昆虫，或者偶尔来块鲜肉。”

“在一个没多少动物的种植园里，哪里能找到肉呢？”

“我们菜单上的主要肉食是新鲜捕获的老鼠。”

“老鼠？你吃老鼠？”大象问道，做了个鬼脸。

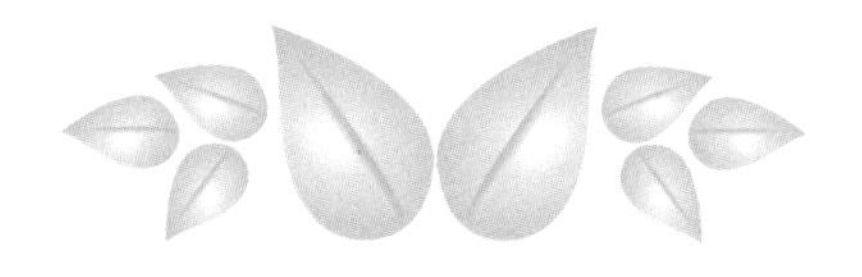

“Nope, more of an omnivore. I do like some insects, or a piece of fresh meat once in a while.”

“Where would you find meat in a plantation, with its very few animals?”

“The main meat item on our menu is freshly caught rat.”

“Rats? You eat rats?” Elephant asks, pulling a face.

“这些老鼠大嚼棕榈果，长得很肥，很好吃！你没试过吗？一旦我们发现一只老鼠，整个队伍就会发起追捕，而老鼠很少能逃脱。”

“喜欢吃老鼠？好吧，我想，各有各的……”

“每天大约10只，仅我们一伙一年就能消灭大约3 000只老鼠。那是怎样的虫害控制啊！你找不到比我们更好、更快、更适合这个工作的了。”

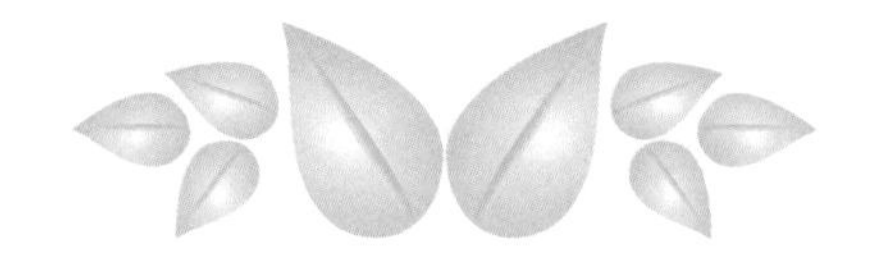

“These rats have grown fat munching on palm fruit and are delicious! Have you never tried them? When we spot one, the whole troop mounts a chase, and the rat seldom escapes.”

“Enjoying eating a rat? Well, I suppose, each to his own…”

“At about ten rats a day, my troop alone kills about three thousand rats a year. How is that for pest control! You won’t find anyone better, swifter or fitter than us for the job.”

大嚼棕榈果，长得很肥

grown fat munching on palm fruit

被诱捕带走。

trapped and taken away.

“种植园主一定会因此爱你的。老鼠不仅破坏收成，而且如果不使用有害毒药，他们根本无法控制它们。”

“可我们还在受到诬蔑和追捕！有的被诱捕带走，与家人分离，永远失去父母或孩子……你现在明白我为什么如此悲伤了吧？”

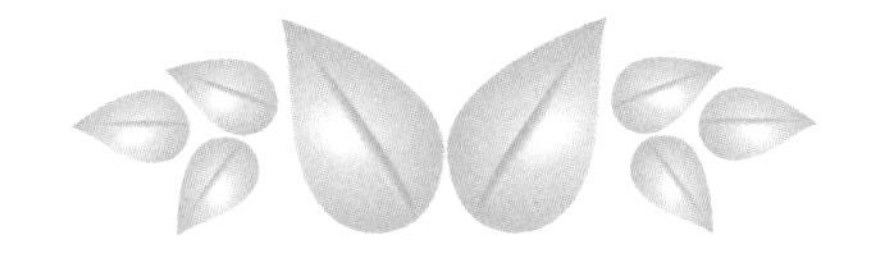

“The plantation owners must love you for that. Not only do rats destroy their harvests, they are impossible to control without using toxins that are bad for everyone.”

“And still we are being vilified and hunted! Or trapped and taken away, separated from our families, losing our parents or young ones, often forever… Do you now understand why I am so sad?”

“我明白了，我也希望人们能意识到，其实你在他们的棕榈园里是很有用的，应该受到欢迎。”

“谢谢。要是他们知道就好了，我多吃一点肉，老鼠就少一些。当然，作为报答，他们可以给我一些甜头！”

……这仅仅是开始！……

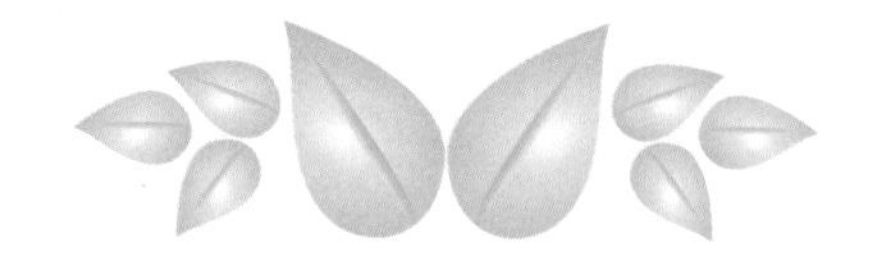

“I do, and I do wish people would realise that you are, in fact, useful to have around, and should be welcomed to their palm plantations.”

“Thank you. If only they knew that, because I enjoy a bit of meat, they have fewer rats. Surely they can thank me by allowing me a few sweet treats!”

... AND IT HAS ONLY JUST BEGUN!...

……这仅仅是开始！……

... AND IT HAS ONLY JUST BEGUN! ...

Did You Know?

你知道吗？

Pig-tailed macaques live in the Indonesian rainforest. Deprived of their traditional feeding grounds, they raid plantations for food. This causes the owners to trap and relocate them, and in the case of returning macaques, shoot them.

短尾猕猴生活在印度尼西亚的热带雨林中。由于失去了传统的觅食场所，它们会偷袭种植园来获取食物。为此种植园主会诱捕它们并把它们送走，如果有猕猴去而复返，就开枪打死它们。

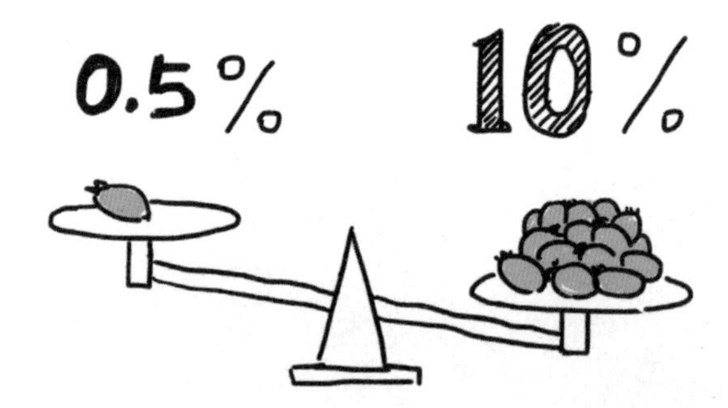

Macaques eat only 0.5% of the palm fruit harvest, and kill rats that can destroy 10% of the harvest. Leaving the macaques to act as rat controllers makes economic sense. It makes sense to apply the principle of “live and let live”.

猕猴吃掉的棕榈果实仅占0.5%，而被其消灭的老鼠却能破坏掉百分之十的收成。让猕猴充当老鼠防控员是划算的。这符合“各不相扰，和谐共存”的原则。

多年来农民一直依赖化学品灭虫，如今越来越多的农民意识到，生物防控更有效、更便宜，而且不会破坏食物链。他们已经意识到使用灭鼠剂会杀死其他以老鼠为食的物种（比如猫头鹰）。

After having relied on chemicals to kill pests for many years, more and more farmers now realise that biological controls are more efficient, cheaper and do not disrupt the food chain. They have realised that using rat poison kills other species (like owls) that eat rats.

棕榈油和棕榈仁油被用在饼干、面包和薯片的加工制作中。它们也被用于巧克力制作，还被添加到牛奶中以增加维生素 A。在印度尼西亚，棕榈油已经取代了传统使用的椰子油，在缅甸已经取代了花生油。

Palm oil and palm kernel oil are used in cookies, bread, and potato chips. It is also used in chocolate, and is added to milk to add Vitamin A. Palm oil has replaced coconut oil traditionally used in Indonesia, and has replaced peanut oil in Myanmar.

Palm oil is extensively used in cosmetics and toiletries soap, shampoo, toothpaste and detergents. It is increasingly being used as a biofuel, which led to a campaign in Europe to "stop putting orangutans in your car's tank".

棕榈油被广泛用于化妆品、香皂、洗发水、牙膏和洗涤剂。随着它越来越多地被用作生物燃料，欧洲发起了一场“不要把猩猩放进你的汽车油箱里”运动。

Between 1990 and 2005, up to 60% of palm oil expansion occurred at the expense of primary tropical rainforests. The establishment of palm oil plantations has been an ecological disaster for endangered wildlife such as the orangutans and tigers.

1990 年至 2005 年间，高达 60% 的棕榈油产业扩张是以牺牲原始热带雨林为代价完成的。建立棕榈种植园对猩猩和老虎等濒危野生动物来说是一场生态灾难。

Large-scale conversion of tropical forests to oil palm plantations impacts on an immense number of plant and animal species. Oil palm production leads to human-wildlife conflict, as large populations of animals are forced into increasingly isolated fragments of natural habitat.

从热带森林到油棕种植园的大规模转变影响到的动植物物种数量巨大。油棕产业导致人类与野生动物之间的冲突，大量的动物被迫进入越来越孤立的小块自然栖息地。

Many national parks have been severely impacted by palm oil production. Illegal palm oil planting has taken over 43% of the Tesso Nilo National Park in Sumatra, a park that was established to provide a safe habitat for the endangered Sumatran Tiger.

许多国家公园受到棕榈油产业的严重影响。苏门答腊岛泰索尼洛国家公园 43% 的土地被非法棕榈种植所占据。该公园是为了给濒危的苏门答腊虎提供一个安全的栖息地而建立的。

Would you rather eat a fresh fruit or a raw rat?

你更愿意吃新鲜的水果还是生老鼠?

How would you control rats without the use of chemicals?

如何在不使用化学品的情况下防控老鼠?

When you are falsely accused, what do you do?

当你被错误指控时，你会怎么做?

Is it a good idea to be vegetarian?

吃素是个好主意吗?

Take a look through the news media to see if you can find reports of people who have been falsely accused, and then later vindicated. What can we, as a society, do for people who have suffered in this way? A mere apology is not enough. Discuss this issue with friends and family members, and ask them how they would approach such an injustice. Keep your focus on finding ways of redressing the situation through restorative justice, rather than any finger pointing or accusations. Look into the measures that can be taken to avoid it happening to others in the future. By educating yourself and others on this issue, it could benefit all.

浏览新闻媒体，看看能否找到人们被诬告后得以平反的报道。社会能为经历这种遭遇的人做些什么？仅仅道歉是不够的。与朋友和家人讨论，问问他们会如何处理这种不公。比起指指点点或谴责，通过恢复性司法纠正这种情况才是重点所在。深入调查，看看能采取哪些措施避免将来他人遇到这类问题。通过这个问题，使自己和他人受到教育并从中受益。

学科知识

Academic Knowledge

生物学	猕猴是食果动物；猕猴是母系社会动物；侏儒象是岛屿侏儒症（为了适应资源贫乏的环境）的结果；褐鼠在三个月大的时候开始繁殖，一年可生育2 000只后代；动物导航的奥秘，从蜣螂到鸽子和家猫。
化　学	用于消灭老鼠的毒药，如砷、铊和金属磷化物，对所有生物都有害；有毒的抗凝剂。
物　理	动物通过磁场导航，并沿地球磁场的南北场线确定它们的位置；其他动物，比如蜣螂，会利用视觉线索，比如银河进行定位；鸽子的喙上有富含铁的细胞；狗通过在重叠的熟悉气味中移动来扩大它们的嗅觉范围，这与手机信号的覆盖依赖于不同信号塔的相互连通很相似。
工程学	灰尘、昆虫、啮齿动物、浣熊和松鼠的捕获技术。
经济学	害兽给农业带来的损失巨大，光老鼠就破坏掉高达20%的收成；在一个“赢家通吃”的市场，最大的赢家占取相当可观的回报，留给剩下的竞争者只有很少的份额；统计学在谈判中建立论据的作用。
伦理学	猕猴被大量用于实验，特别是在神经科学和视觉感知领域；仅仅为追求经济利益而破坏热带雨林以提高植物油产量；单一种植业所付出的代价，因其单一性导致害兽滋生以及对杀虫剂的需求；野生动物和原住民自古以来的土地占有权；缺乏理解、不愿花时间去理解导致公正失衡。
历　史	侏儒象包括史前的猛犸象；在18世纪，孩子们更喜欢捕老鼠而不是扫烟囱；在维多利亚时代，捕鼠是一项官方任务。
地　理	侏儒象曾在撒丁岛、西西里岛、马耳他和克里特岛一带活动；加里曼丹（印度尼西亚）棕榈种植中心；猕猴生活在亚洲和北非。
数　学	运用空间动力学（数学模型，使用复杂的计算机模拟）呈现捕食者如何在时间和空间中捕猎，更有效率意味着在更短的时间内捕获相同数量的猎物。
生活方式	严格的素食主义者和偏好素食的人拒绝动物产品的程度有所不同；老鼠的繁殖与生活方式有关；生肉含有更多的维生素B，在法国、比利时、意大利受到推崇；宗教禁止食用特定的动物产品。
社会学	千禧一代日渐增加的孤独感对公共健康产生影响；老鼠在人群中传播疾病，包括鼠疫；在危机时刻能够依靠家庭的重要性，让他们在逆境中也能成长。
心理学	挫折是在满足需求或达到目标的过程中遇到反对而产生的情绪反应；悲伤是由于丧失、绝望、悲伤或不利的感觉而引起的情感上的痛苦；孤独是由于感到被孤立而产生的不愉快的情绪。
系统论	老鼠导致了哺乳动物、鸟类和爬行动物的灭绝，尤其是岛屿上的。

情感智慧

Emotional Intelligence

佚儒象

侏儒象问猕猴伤心的原因，表达了对她的同情。他对同伴的痛苦表示关心，并推测使她悲伤的原因。不能立即得到明确答案时，他耐心地坚持用他的方式提问。猕猴与他分享了悲伤的原因，他听取了解释，并问了更多问题试图完全了解情况。他对猕猴吃老鼠感到惊讶，并无法掩饰他的厌恶之情。尽管如此，他还是承认自己不喜欢的食物别人有可能会喜欢。他希望种植园主能更好地了解猕猴在防治害兽方面的作用，这可能会使他们受到棕榈园的欢迎。

猕　猴

猕猴非常悲伤。起初，处在极度痛苦中的她无法谈及原因。后来她讲述了森林遭破坏的历史和对她及家庭的影响。尽管遭受挫折，他们仍然在尚存的雨林地区顽强生存下来，学会满足现状，确保没有挨饿，家庭兴旺。她不喜欢被称为害兽并受到欺凌，她认为自己受到种植园主的诽谤和不公对待。她坚持认为猕猴有权公平地分到土地上的果实。侏儒象的鼓励给了她信心，她开始更多地分享她的生活方式和饮食偏好。她解释了猕猴如何集体协作猎捕老鼠，并分享数据以提供透明度。

艺术

The Arts

让我们试着画出不同的面部表情。比如，你会怎样表现悲伤？一个简单的方法就是画一个圆圈，用两个点画眼睛，用半月形画嘴巴。自由发挥创造力，用自己的笔触和手法画出不同的简笔画。嘴巴可以是不同的大小和形状，眼睛或大或小，或斜或闭。然后画上眼泪，或者通过改变肤色来增加一点深度。你正在设计自己的表情包！你能添加一些表达失望的内容吗？设计一些能让我们发笑的东西如何？

思维拓展

Systems: Making the Connections

现象与其背后的社会环境因素密不可分。猕猴要求公平分配栖居地的果实。随着棕榈种植园开发，数百万公顷的雨林被砍伐。猕猴和其他生物（如侏儒象）共享的自然栖息地遭到破坏。现代资本主义所有的关注都在经济效益上。猕猴已经适应了自然栖息地的丧失，并接受了新的居住条件。然而，种植园主只关注棕榈种植园的生产力，猕猴的积极作用未得到认可。他们认为猕猴吃棕榈果会减少产量，对猕猴进行猎捕和驱逐，却没有考虑到被消灭掉的老鼠数量远高于被吃掉的果子数量。种植园主需要了解因种植棕榈而改变的生态系统，对所有生物的生存和利益都加以考虑。鼠害防治的典型办法是使用有毒化学药剂，这会影响整个生态系统和产品质量。雨林已遭到破坏，短时间内无法得到修复。面对这残酷的现实，我们必须利用现有资源，找到一个所有人都能生存下去的模式。我们需要种植园主的加入和共识，共同致力于人类与野生动物互惠的解决方案。

动手能力

Capacity to Implement

你是否在家附近遇到过鸟、老鼠、蛇和浣熊这些野生动物？鸽子在窗台上排泄的粪便，邋遢的浣熊来吃猫狗的食物，松鼠想要靠近鸡或鸟的喂食器，面对这些情况我们该怎么做？与其杀害，不如放归野外。不要直接接触这些动物，因为它们可能携带寄生虫或其他有害病菌，防护装备也可以防止你被咬伤或划伤。

如果动物被送走后再度返回，或者你正在被老鼠侵扰，想一想是什么导致它们大量存在。你要做的第一步应该是确保环境整洁，这样它们就不容易获取食物。如果没有东西可吃，大部分动物和鸟类就不会进入你的房子或花园，你也就不用诱捕它们了！

故事灵感来自

This Fable Is Inspired by

托丽·赫里奇

Tori Herridge

托丽·赫里奇博士于 2002 年毕业于伦敦大学学院，获生物学学位。在伦敦帝国理工学院取得硕士学位后，她继续攻读博士学位，研究地中海岛屿上已灭绝的侏儒象。在伦敦自然历史博物馆工作期间，赫里奇博士通过讲座和电视纪录片普及科学知识。她撰写了关于克隆猛犸象的伦理学，以及拯救濒危大象的重要性等文章。她是“泥刀使者”的联合创始人，该组织致力于展示开拓型女性在古生物学、地质学和考古学领域的贡献。

图书在版编目(CIP)数据

冈特生态童书.第七辑：全36册：汉英对照 /
(比)冈特·鲍利著;(哥伦)凯瑟琳娜·巴赫绘；
何家振等译.—上海：上海远东出版社，2020
ISBN 978-7-5476-1671-0

Ⅰ.①冈… Ⅱ.①冈… ②凯… ③何… Ⅲ.①生态
环境-环境保护-儿童读物—汉英 Ⅳ.①X171.1-49

中国版本图书馆CIP数据核字(2020)第236911号

策　　划 张　蓉
责任编辑 程云琦
助理编辑 刘思敏
封面设计 魏　来李　廉

冈特生态童书
我做的是好事!
[比]冈特·鲍利　著
[哥伦]凯瑟琳娜·巴赫　绘
颜莹莹　译